YOUR KNOWLEDGE HAS VALUE

Nazim Nariman

Vibration and Mode Shapes Analysis of Cable Stayed Bridges Considering Different Structural Parameters

GRIN Verlag

Bibliografische Information der Deutschen Nationalbibliothek:

Die Deutsche Bibliothek verzeichnet diese Publikation in der Deutschen National-
bibliografie; detaillierte bibliografische Daten sind im Internet über http://dnb.d-
nb.de/ abrufbar.

Imprint:

Copyright © 2014 GRIN Verlag GmbH
Druck und Bindung: Books on Demand GmbH, Norderstedt Germany
ISBN: 978-3-656-58878-8

This book at GRIN:

http://www.grin.com/en/e-book/267379/vibration-and-mode-shapes-analysis-of-
cable-stayed-bridges-considering

Title

Vibration and Mode Shapes Analysis of Cable Stayed Bridges
Considering Different Structural Parameters

Author

Nazim Abdul Nariman

PhD candidate - Institute of Structural Mechanic

Bauhaus Universitat Weimar – Germany

Abstract

The vibration characteristic of a cable stayed bridges structure is the main axis of the study in this paper, many structural parameters are used to simulate and determine the effect of vibration on the structural performance by identifying the natural frequencies of the system and the mode shapes that can occur in the real structure. Modeling the stay cables with three famous styles of arrangements such as Harp, Semi Harp and Fan styles, and assigning roller, hinged and fixed boundary conditions on the deck support of the cable stayed bridge, in addition to using two design cases of the girders and pylons dimensions in the global structure for that purpose. Through the use of ABAQUS finite element analysis, the models were generated for each mentioned cases and the results of the frequency linear perturbation step of 10 mode shapes were determined through the simulation of the deformed shapes and the determined values of the natural frequencies of each mode for each case of interest. It was seen that the roller boundary condition was much prone to the early vibration and the stay cables of the direction near to the roller support were vibrated and stressed much more than the other direction compared with the hinged and fixed boundary conditions, and the mode shapes 7, 8, 9 and 10 were the most vibrated cases for all the boundary conditions without any distinction. The week design of the girders and the pylons has the great effect on the vibration of the stay cables, pylons and deck of the structure especially near the roller support direction due to the early vibration of the case of roller support, so the use of cross ties and damping between the stay cables and the girders are very important in the cases of significant vibrations which affect the performance of the cable stayed bridges.

Keywords

Natural Frequency, Vibration, Mode shapes, Eigenvalues, Cross Ties and Damping

Introduction

Construction of long-span Bridge has been very active worldwide in the past few decades. Some of the cable-stayed bridges exceeding 1000 m, such as Stonecutters, Sutong in China and Russky Bridge in Russia with central span of 1104 m were completed in the beginning of 21st century. A number of long-span bridges are now under construction in China and Korea, and the plans to build super long-span bridges in other parts of the world are also being discussed. As bridge spans get longer and pylons get taller, they become more flexible and prone to vibration. Flexible structures tend to vibrate under dynamic loading such as wind, earthquake, vehicle movement, and human motion. Vibration can have several levels of consequence, from a potentially hazardous effect such as causing immediate structural failure to a more subtle but more extended effect such as structural fatigues. In addition to that, vibration can also affect user safety and comfort and limit bridge serviceability. In the past few decades, extensive research and development has been carried out to understand the mechanisms behind bridge vibration and to reduce undesirable vibration effects through various countermeasures. Results of these research and developments have been adopted in bridge design codes and put into practices by specifying methodologies and guidelines for countermeasures, and by introducing new structural elements or devices as vibration control. Types of vibration commonly observed on the long-span bridge are discussed from structure engineering viewpoints. Some vibration mechanisms are now well understood, while some others still require further studies to achieve complete understanding. Surveys on the phenomena associated with the type of vibration reported in many long-spa n bridges are also presented as well as engineering solutions adopted as countermeasures. The progressive lengthening of the spans of cable-stayed bridges has strongly increased the importance of understanding the dynamic behavior of cable-stayed bridges. As slender, flexible and typically three-dimensional structures, cable-stayed bridges pose specific vibration problems under dynamic loading such as earthquake, wind, rain and traffic. Cables may lead to complex deck-cable and tower-cable coupling vibrations. Previous studies have revealed the importance of cable vibration in dynamics of cable-stayed bridges. It is found that the cable vibration can have a significant influence on the modal characteristics and seismic response of cable-stayed bridges when the first frequencies of cable local modes overlap with the first few frequencies of bridge deck/tower. Therefore, developing an accurate finite element model accounting for the interaction of cable vibration with bridge vibration is essential to understanding the dynamic behavior of a cable-stayed bridge.

Cable-stayed bridges usually show long fundamental periods, aspect that influences their dynamic behavior. However, their flexibility and dynamics characteristics depend on several parameters such as the main span length, stay cable layout and support conditions. First vibration modes are very long, mainly related to deck modes. They are followed by cable vibration modes or tower modes that can be coupled with the deck depending on the support conditions. An exact modeling for the deck and cables can be very important for a precise dynamic analysis, being necessary an adequate assessment of the natural frequencies and modal shapes.

Objective

The objective of this research work is to search the vibration characteristics of cable stayed bridge models so that to:

1- Discover the effect of the boundary condition on the vibration of the stay cables, deck and the pylons and to find the suitable boundary condition for each case as a result to define the regions that are much prone to early vibration and to control this mitigation in the cable stayed bridges.

2- Identify the relation between the stay cables styles and the early vibration by simulation of mode shapes as a result to get the natural frequencies of the system to increase their values by adding cross ties on the stay cables in the areas of significant detected vibrations.

3- Retrofitting of damaged cable stayed bridges that have been poorly designed through simulation of the structure vibration with negative (high vibration and deflection) of girders and pylons which will prepare a safe structure against vibration.

Natural Frequencies and Mode Shapes

In typical structural or mechanical systems, there are many multiple or nearly equal natural frequencies due to their structural symmetries or certain reasons. In this case, since eigenspace spanned by the mode shapes corresponding to the multiple natural frequencies are degenerate, any linear combination of mode shapes can be a mode shape. For the mode shape derivative to be found, the adjacent mode shapes which lie adjacent to the multiplicity of multiple natural frequency distinct mode shapes appearing when a design parameter varies must be calculated first. To do so, the approximate mode shapes could be varied continuously by varying the design parameter. For the real symmetric case, a generalization of Nelson's method was obtained by Ojalvo and amended by Mills-Curren and Dailey. Dailey's method is an exact analytical method for calculating mode shape derivatives. This method only requires knowledge of the eigenpair with multiple eigenvalues, however, the method is lengthy and complicated for finding mode shape derivatives and clumsy for programming. Dailey's method is extremely complicated for calculating the sensitivity of eigenvectors of multiple eigenvalues in the case of the damped systems. When a natural frequency has multiplicity m and a design parameter is perturbed, the corresponding mode shapes may split into as many as m distinct mode shapes. For derivatives of the mode shapes to be responsible, the mode shapes must be laid adjacent to the m distinct mode shapes that appear when a design parameter varies. The eigenvalue problem of a damped system can be expressed as.

$$(\lambda^2 M + \lambda C + K)\phi = 0 \tag{1}$$

Where M, C and K are the matrices of mass, damping and stiffness, respectively and definite or these are order n symmetric matrices. M is positive definite and K positive semi-positive definite. The first step in finding derivatives of mode shapes of multiple eigenvalues is to find corresponding adjacent mode shapes Suppose that all eigenpairs are known and multiplicity of the eigenvalue lm is m. Define the following eigenvalue problem where Fm is the matrix of eigenvectors correspond to the multiple eigenvalue, hence, its order (n*m).

$$M\Phi_m \Lambda_m^2 + C\Phi_m \Lambda_m + K\Phi_m = 0 \tag{2}$$

Where the eigenvalue and eigenvectors are:

$$\Lambda_m = \lambda_m I_m \tag{3}$$

$$\Phi_m = \left[\phi_{i+1}\phi_{i+2}\cdots\phi_{i+m}\right] \tag{4}$$

Im is the identity matrix of order m and lm is the eigenvalue of multiplicity m for the eigenspace.

Modal Analysis Theory

The vibration characteristics decide the dynamic response characteristics. Consequently the modal analysis of the cable-stayed bridge is essential to study the dynamic behavior of cable-stayed bridge. The structure of the cable-stayed bridge is complex, and the cables are flexible, lightweight, low damping, etc. All these features above make the mode of the cable-stayed bridge a practical engineering issue of worth attention. The structure of the cable-stayed bridge is a system with continuous distribution of mass and stiffness. It should be divided into finite elements with limited DOF. Because of the complexity of cable-stayed bridge structure, the result of three-dimensional FEM analysis is more comprehensive and more reliable than the result of the traditional empirical formula. It can be assumed that the structural of the cable-stayed bridge has N DOF. The dynamic equilibrium equation of the model can be list as:

$$[M]\{\ddot{U}\}+[C]\{\dot{U}\}+[K]\{U\} = \{P(t)\} \tag{5}$$

Where [M] = mass matrix of the architecture, [C] = damping matrix of the architecture, [K] = stiffness matrix of the architecture,
{U} = displacement vector of each node, {Ŭ} = velocity vector of each node,
{Ű} = acceleration vector of each node. Ignoring the resistance, we can get a dynamic equilibrium equation:

$$[M]\{\ddot{U}\}+[K]\{U\} = 0 \tag{6}$$

take $[U(t)] = \{\Phi\}\sin\omega t$ \hfill (7)

and then solve the differential equations. We can obtain:

$$\left(\left[K\right]-\omega^2\left[M\right]\right)\{\Phi\}=0 \tag{8}$$

We can get natural frequencies of the system ωi (i=1, 2,...,N) and the vibration modals of the structure $\{\Phi i\}$ {i=1, 2,...,N} from this equation.

The initial force in the cables and the gravity of the bridge has little effect on the dynamic characteristics of the structure. Therefore, the initial force in the cable and gravity of the bridge can be ignored when modal analysis carried out. By constraining the corresponding nodes degrees, the boundary conditions of the model can be defined.

Equations of Motion

The vibration of CSBs due to vehicular loading was analyzed via FEM with beam and cable elements. The substructure method proposed by Nagai was applied to consider the local vibration of cables. Each simulated cable was divided into $n-1$ Inter linking truss elements **Fig. 1**.

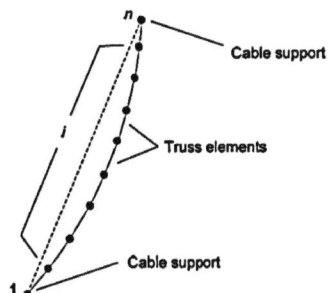

Fig. 1 Finite Element Model of the Cable

To improve the computational efficiency of the dynamic analysis of CSBs, the local vibrations of cables were considered using a substructure method. According to the method of superposition, the vibrations of the cable are composed of vibrations of cable supports and several modes. The accuracy of the method depends on the quantity of modes adopted in calculation. For flexible structures such as cables, the vibration is composed primarily of the components of low orders of modes. The equation of motion of the free vibrations of each cable can be expressed as:

$$\begin{bmatrix} m_{11} & m_{1i} & 0 \\ m_{i1} & m_{ii} & m_{in} \\ 0 & m_{ni} & m_{nn} \end{bmatrix} \begin{bmatrix} \ddot{d}_1 \\ \ddot{d}_i \\ \ddot{d}_n \end{bmatrix} + \begin{bmatrix} k_{11} & k_{1i} & 0 \\ k_{i1} & k_{ii} & k_{in} \\ 0 & k_{n1} & k_{nn} \end{bmatrix} \begin{bmatrix} d_1 \\ d_i \\ d_n \end{bmatrix} = \begin{bmatrix} 0 \\ 0 \\ 0 \end{bmatrix} \tag{9}$$

where *mij* are the sub-matrices of the mass matrix, *kij* are the sub-matrices of the stiffness matrix, *d*1, *dn* are the displacement vectors at cable supports and *di* represents the displacement vector of the inner nodes of the cable. According to the superposition method **Fig.2**, *di* can be separated into two parts, the movement of cable supports and modal motions. In this study, the modal motion of a cable is assumed to be a combination of several lower-order modes of the cable with fixed supports. Thus, *di* is approximated as

$$d_i \approx T_1 d_1 + T_2 d_n + \Phi q \tag{10}$$

Where

$$\begin{cases} T_1 = -k_{ii}^{-1} k_{i1}, \\ T_2 = -k_{ii}^{-1} k_{in}, \end{cases} \tag{11}$$

and *q* is the generalized coordinates vector. Φ is the modal matrix of a cable with two fixed ends, and is obtained by solving the eigenvalue problem from equation (7)

$$\Phi = \begin{bmatrix} \phi_1 & \phi_2 & \cdots & \phi_m \end{bmatrix} \tag{12}$$

where *φi* is the *i*th natural mode and *m* is the quantity of modes adopted in the vibration calculation.

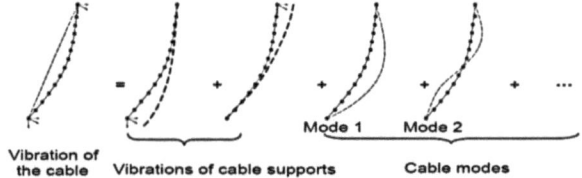

Vibration of the cable   Vibrations of cable supports   Cable modes

Fig. 2 Method for calculating the local vibration of cables

Hence,

$$\tilde{M}_c \begin{bmatrix} \ddot{d}_1 \\ \ddot{d}_n \\ \ddot{q} \end{bmatrix} + \tilde{K}_c \begin{bmatrix} d_1 \\ d_n \\ q \end{bmatrix} = 0 \tag{13}$$

Where

$$\begin{cases} \widetilde{M}_c = T_s^T M_c T_s, \\ \widetilde{K}_c = T_s^T K_c T_s, \end{cases} \tag{14}$$

where Mc and Kc are the mass and stiffness matrices, respectively, M`c and K`c are the equivalent mass and stiffness matrices, respectively after freedom quantity reduction, and

$$T_s = \begin{bmatrix} I & 0 & 0 \\ T_1 & T_2 & \Phi \\ 0 & I & 0 \end{bmatrix} \tag{15}$$

where I is the unit matrix.

Cable Vibrations

Cables are essential components of long-span bridges and they are prone to vibration because of their higher flexibility, small mass and low mechanical damping. Stay cables of cable-stayed bridge is generally very flexible compared to other bridge components, and therefore prone to vibrate. In long-span cable-stayed bridges, stay cable can be as long as 500 m, resulting in low natural frequencies of about 0,2 to 0,3 Hz in the lowest mode. A survey on inherent damping of stay cables of cable-stayed bridges in Japan shows that structural damping could reach as low as 0,1% critical damping ratio. Large cable oscillations may cause damage to the anchorage and cable fatigue.

Frequent and excessive cable vibrations will cause cable fatigue failure at its anchorage. There are lots of cable fault in the past. For example, it was found on Maracaibo Bridge in Venezuela that more than 500 steel wires were damaged. A year later, another 3 cables were totally failed. In the case of Jinan Yellow River Bridge in China, all the cables were replaced after 13 years in service because of fatigue failure. Stay cables account at least 15% of the total cost of a cable-stayed bridge project. Therefore, reducing the probability of stay cable failure caused by vibration is very important not only in terms of structural safety, but for economy as well.

Free Vibration Response

Solutions for the free vibrations of a cable that is fixed at both ends were given by Irvine. Their results can easily be derived from the equation of static stiffness of a taut wire as shown in **Fig. 3** below:

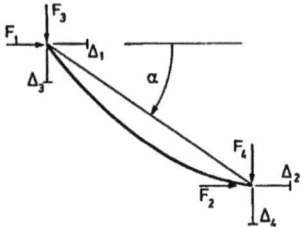

Fig. 3 Global Force and displacement quantities of a cable

Elastic part of the static stiffness of taut wire is

$$K_{11}^{t,e} = \frac{EA}{l} \cos^2 \alpha \tag{16}$$

The static stiffness of taut wire is

$$K_{11}^{t} = K_{11}^{t,e} \left(1 + \frac{T\theta}{EA} \tan^2 \alpha \right) \tag{17}$$

The stiffness function that is described by this equation becomes infinite for certain values of Ωc, in the case of real Ωc these values coincide with the dimensionless natural frequencies of un-damped cable suspended from rigid end supports. This condition leads to the frequency equations:

$$\tan \frac{\Omega}{2} = 0 \tag{18}$$

$$\tan \frac{\Omega}{2} - \frac{\Omega}{2} + \frac{4}{\lambda^2} \left(\frac{\Omega}{2} \right)^3 = 0 \tag{19}$$

Where the dimensionless natural frequency Ω is defined as

$$\Omega = \omega l \sqrt{\frac{m}{T\theta}} \tag{20}$$

And in which w is the natural circular frequency. And the solution for equation (18) is:

$$\Omega_n = 2n\pi \tag{21}$$

Where $n = 1, 2, 3......$

and corresponds to the natural frequencies of the anti-symmetric in-plane modes. The roots of the transcendental equation **(20)** correspond to the natural frequencies associated with symmetric in-plane modes.

Cable Elements

The major nonlinearities of the inclined cables result from the sag effect. The most accepted and convenient way to account for this effect is to use a straight member, instead of a curved one, with the equivalent modulus of elasticity. This form was first proposed by Ernst and has been verified by several tests. The equivalent cable modulus of elasticity can be given by:

$$Eeq = \frac{E}{1 + \left[\dfrac{(w_{cL})^2 AE}{12T^3}\right]} \tag{22}$$

Where E is modulus of elasticity, W_{cL} is cable weight per unit length, A is cable area, L is horizontal projected length, T is cable tension. Then, the stiffness matrix of the cable element in the local coordinates is expressed by the following:

$$[K_E] = \frac{E_{eq}A}{l}\begin{bmatrix} 1 & -1 \\ -1 & 1 \end{bmatrix} \tag{23}$$

And l is the cable length.

Cross Ties

Using cross ties is one of the considerable countermeasures for vibration mitigation of long stay cables; however, the analysis method for design of cross ties has not been established yet. One mechanism of the cross ties is to increase the frequencies of stay cables, another mechanism is to increase the damping, **Fig. 4** shows the cross ties.

The number, distribution, tensioning method and initial tension of cross ties were discussed by Zhou and then he suggested two types of cross ties, stiff and flexible types.

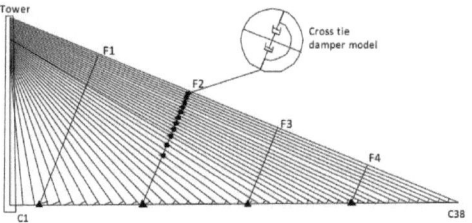

Fig. 4 FEM model of cross ties on stay cables

Finite element models of vibration

A main model of a cable stayed bridge is created in ABAQUS commercial software with 324 m length using three dimensional Beam elements to model the Pylon, Deck and Stay Cables. The bridge is modeled using steel box cross section for the girders of the deck With 6 M length, 2 M width and 0.3 M thickness, and mass density 8500 K/M3, modulus of elasticity 200 MPa and poisons ratio 0.3.

The pylons are 30 M height over the deck and 40 M under the deck through the sea water, and are stated in symmetric positions. They are made of high strength reinforced concrete of mass density 2500 Kg/M3, modulus of elasticity 30 Mpa and poisons ratio 0.2 with especial techniques for the stay cables to fix in the designated positions along the pylon height, the cross section of the pylons is rectangular of 6 M length, 3 M width.

The stay cables are used to support the deck and transfer the loads and stresses to the pylons in many arrangements as it shown in **Fig.5**, **Fig.6** and **Fig.7**. The stay cables are made using parallel wire strand stays with mass density of 8500 Kg/M3, modulus of elasticity 200 Mpa and poisons ratio 0.3. The number of stay cables is 80 stay cables distributed on the pylons symmetrically.

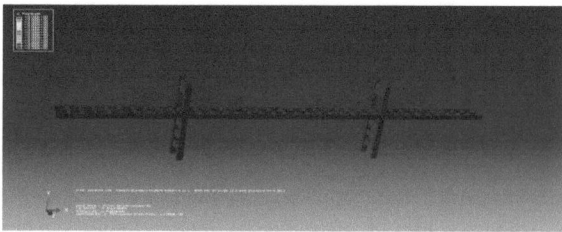

Fig. 5 Finite Element Model (Harp type) Rendered View

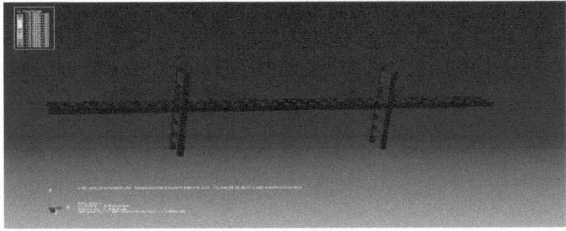

Fig. 6 Finite Element Model (Semi-harp type) Rendered View

Fig. 7 Finite Element Model (Fan type) Rendered view

General Static Analysis

A Linear perturbation frequency step was created with permission of (Nlgeom) aspect which means showing the non-Linear geometrical effects of the large displacement, so the initial stress stiffness effect of any loading applied in a prior step will be taken into account and the large strains would be taken in consideration. The numbers of eigenvalues requested in this step were 10 values which mean 10 mode shapes of the model response due to vibration induced from the natural frequencies of the system.

Boundary Conditions of the Model

The deck support in the left side is fixed to prevent any translation of the deck in the X and Y directions, in addition to preventing rotation about Z axis, But the support in the right side permits only rotation about Z axis without any translation in X and Y directions, this step was adopted to stabilize the bridge against any lateral displacement of the deck and to absorb any excited vibration due to the loads assigned on the cable stayed bridge. The pylon supports are all fixed and prevent any translation and rotation as shown in **Fig.8**.

Fig. 8 Boundary Conditions of the Deck supports and Pylons support

Seeding and Meshing Step

The meshing process for the model part needs seeding operation, so a global seeding size of 10 was selected to mesh the model part with a standard Linear Beam Element **B31** for all the cases of cable stayed models compromising the Fan, Semi-harp and Harp arrangement cases for the stay cables optimization. **Fig.9** shows the meshing of the cable stayed model with a Fan arrangement of the stay cables.

Fig. 9 Seeding of the part and Meshing

Vibration Analysis

Many structural parameters is considered for the vibration analysis which are shown in details in the following section.

Boundary Condition Effect on Vibration

Models of cables stayed bridges with different boundary conditions for the right support of the deck are generated for each case of stay cables arrangements, and the results of the frequency step are determined for eigenvalues and natural frequencies of the system for each case as shown in the following **Tables 1,2,3,4,5,6,7,8** and **9** that belong to 10 mode shapes.

Harp Style Model

The results of 10 mode shapes with their natural frequencies associated with the Harp style are shown in the following **Table 1, 2** and **3**.

Table.1 Vibration analysis of Harp Model with Roller Support

Mode Shape	Eigenvalue	Natural Frequency cycle/time
1	5.1098	0.35977
2	5.1165	0.36
3	5.1204	0.36014
4	5.1213	0.36017
5	5.1213	0.36017
6	5.1214	0.36017
7	5.1222	0.3602
8	5.123	0.36023
9	5.1232	0.36024
10	5.1233	0.36024

Fig. 10 Mode Shape 1 Harp Style Roller Support

The mode shape 1 of the cable stayed bridge model of Harp style shows the vibration of the stay cables in the direction near the roller support with significant amounts of stress and deflection in the longest four stay cables in addition to very small vibration and stress in the stay cables in the other direction.

Table.2 Vibration analysis of Harp Model with Hinged Support

Mode Shape	Eigenvalue	Natural Frequency cycle/time
1	5.1156	0.35997
2	5.1194	0.36011
3	5.1211	0.36017
4	5.1213	0.36017
5	5.1213	0.36017
6	5.1214	0.36018
7	5.1222	0.3602
8	5.123	0.36023
9	5.1232	0.36024
10	5.1234	0.36024

Fig. 11 Mode Shape 1 Harp Style Hinged Support

The mode shape 1 of the cable stayed bridge model of Harp style shows the vibration of the stay cables in both directions with significant amounts of stress and deflection in the longest eight stay cables.

Table.3 Vibration analysis of Harp Model with Fixed Support

Mode Shape	Eigenvalue	Natural Frequency cycle/time
1	5.1156	0.35997
2	5.1194	0.36011
3	5.1211	0.36017
4	5.1213	0.36017
5	5.1213	0.36017
6	5.1214	0.36018
7	5.1224	0.36021
8	5.1231	0.36024
9	5.1232	0.36024
10	5.1234	0.36024

Fig. 12 Mode Shape 1 Harp Style Fixed Support

The mode shape 1 of the cable stayed bridge model of Harp style shows the vibration of the stay cables in both directions with significant amounts of stress and deflection in the longest eight stay cables the same as in the hinged support case.

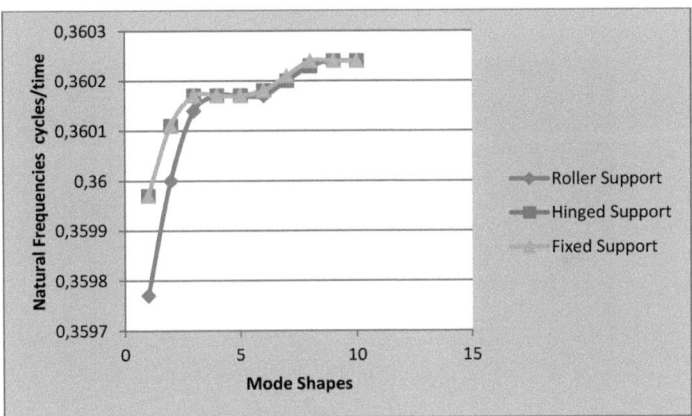

Fig.13 Natural Frequencies and Boundary Conditions Harp Style

Early vibration for the three first mode shapes can be seen in the roller support case compared with hinged and fixed cases which means that the natural frequencies of the system in the model of roller support boundary condition is lower than those for the hinged and fixed ones, but the rest seven mode shapes of all cases are with similar natural frequencies almost.

Semi Harp Style Model

The results of 10 mode shapes with their natural frequencies associated with the Semi Harp style are shown in the following **Table 4, 5** and **6**.

Table.4 Vibration analysis of Semi Harp Model with Roller Support

Mode Shape	Eigenvalue	Natural Frequency cycle/time
1	5.1098	0.35977
2	5.1165	0.36
3	5.1204	0.36014
4	5.1213	0.36017
5	5.1213	0.36017
6	5.1213	0.36017
7	5.1222	0.3602
8	5.123	0.36023
9	5.1232	0.36024
10	5.1233	0.36024

Fig. 14 Mode Shape 1 Semi Harp Style Roller Support

The mode shape 1 of the cable stayed bridge model of Semi Harp style shows the vibration of the stay cables in the direction near the roller support with significant amounts of stress and deflection in the longest four stay cables.

Table.5 Vibration analysis of Semi Harp Model with Hinged Support

Mode Shape	Eigenvalue	Natural Frequency cycle/time
1	5.1156	0.35997
2	5.1195	0.36011
3	5.1211	0.36017
4	5.1213	0.36017
5	5.1213	0.36017
6	5.1214	0.36018
7	5.1222	0.3602
8	5.123	0.36023
9	5.1232	0.36024
10	5.1233	0.36024

Fig. 15 Mode Shape 1 Semi Harp Style Hinged Support

The mode shape 1 of the cable stayed bridge model of Semi Harp style shows the vibration of the stay cables in both directions with significant amounts of stress and deflection in the longest eight stay cables.

Table.6 Vibration analysis of Semi Harp Model with Fixed Support

Mode Shape	Eigenvalue	Natural Frequency cycle/time
1	5.1157	0.35997
2	5.1195	0.36011
3	5.1211	0.36017
4	5.1213	0.36017
5	5.1213	0.36017
6	5.1214	0.36018
7	5.1224	0.36021
8	5.1231	0.36024
9	5.1232	0.36024
10	5.1234	0.36024

Fig. 16 Mode Shape 1 Semi Harp Style Fixed Support

The mode shape 1 of the cable stayed bridge model of Semi Harp style shows the vibration of the stay cables in both directions with significant amounts of stress and deflection in the longest eight stay cables the same as in the hinged support case.

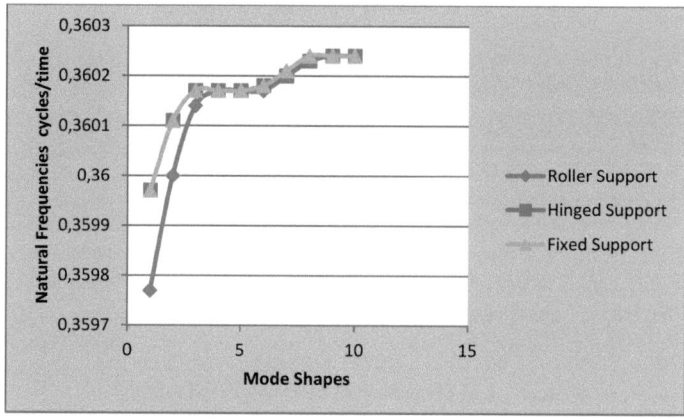

Fig.17 Natural Frequencies and Boundary Conditions Semi Harp Style

Early vibration for the three first mode shapes can be seen in the roller support case compared with hinged and fixed cases which means that the natural frequencies of the system in the model of roller support boundary condition is lower than those for the hinged and fixed ones, but the rest seven mode shapes of all cases are with similar natural frequencies almost.

Fan Style Model

The results of 10 mode shapes with their natural frequencies associated with the Fan style are shown in the following **Table7, 8** and **9**.

Table.7 Vibration analysis of Fan Model with Roller Support

Mode Shape	Eigenvalue	Natural Frequency cycle/time
1	5.1096	0.35976
2	5.1167	0.36001
3	5.1205	0.36014
4	5.1213	0.36017
5	5.1213	0.36017
6	5.1213	0.36017
7	5.122	0.3602
8	5.1228	0.36023
9	5.1231	0.36024
10	5.1233	0.36024

Fig. 18 Mode Shape 1 Fan Style Roller Support

The mode shape 1 of the cable stayed bridge model of Fan style shows the vibration of the stay cables in the direction near the roller support with significant amounts of stress and deflection in the longest four stay cables in addition to very small vibration and stress in the stay cables in the other direction.

Table.8 Vibration analysis of Fan Model with Hinged Support

Mode Shape	Eigenvalue	Natural Frequency cycle/time
1	5.1158	0.35998
2	5.1196	0.36011
3	5.1211	0.36017
4	5.1213	0.36017
5	5.1213	0.36017
6	5.1214	0.36018
7	5.122	0.3602
8	5.1229	0.36023
9	5.1231	0.36024
10	5.1233	0.36024

Fig. 19 Mode Shape 1 Fan Style Hinged Support

The mode shape 1 of the cable stayed bridge model of Fan style shows the vibration of the stay cables in both directions with significant amounts of stress and deflection in the longest eight stay cables.

Table.9 Vibration analysis of Fan Model with Fixed Support

Mode Shape	Eigenvalue	Natural Frequency cycle/time
1	5.1158	0.35998
2	5.1197	0.36012
3	5.1211	0.36017
4	5.1213	0.36017
5	5.1213	0.36017
6	5.1214	0.36018
7	5.1223	0.36021
8	5.123	0.36023
9	5.1231	0.36024
10	5.1233	0.36024

Fig. 20 Mode Shape 1 Fan Style Fixed Support

The mode shape 1 of the cable stayed bridge model of Fan style shows the vibration of the stay cables in both directions with significant amounts of stress and deflection in the longest eight stay cables the same as in the hinged support case.

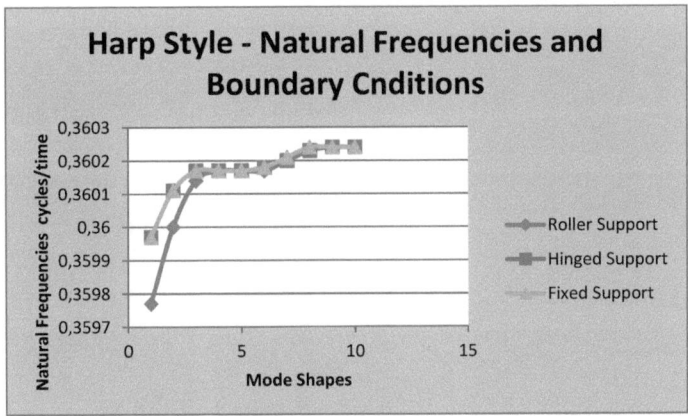

Fig. 21 Natural Frequencies and Boundary Conditions Fan Style

Early vibration for the three first mode shapes can be seen in the roller support case compared with hinged and fixed cases which means that the natural frequencies of the system in the model of roller support boundary condition is lower than those for the hinged and fixed ones, but the rest seven mode shapes of all cases are with similar natural frequencies almost.

Stay Cables Styles Effect on Vibration

The results of the 10 mode shapes of Harp, Semi Harp and Fan styles are shown in **Fig.22**, for analysis of mode shapes, the roller support case has been taken in consideration because of the early vibration in the system. It is obvious that there is difference in vibration and mode shapes for all mode shapes for all cases except the mode shapes 7, 8, 9 and 10 which are similar in response related with high amount of vibration in the longest stay cables.

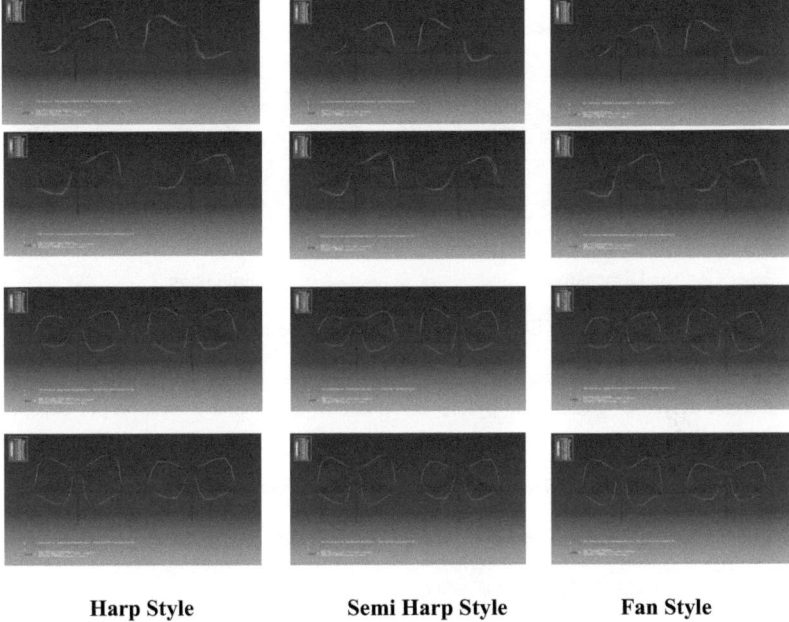

| Harp Style | Semi Harp Style | Fan Style |

Fig. 22 10 Mode Shapes for Harp, Semi Harp and Fan Styles (Roller Support)

Mitigation of stay cables vibration exists in all cases, need to minimize this aspect arises by taking technical application of cross ties in the areas on the stay cables which are prone to high vibration and presence of damping also is a practical solution beside the cross ties for the stability of the structure against vibration.

Pylons and Girders Effects on Vibration

By taking the first mode shape of vibration for each case of the stay cables styles and changing the dimension of the Pylons and the Girders which have great effects on the vibration of the structure especially on the stay cables in addition to the deck and the pylons in the direction near the roller support for Harp, Semi Harp and Fan styles cases especially in the weak design of the girders and pylons which dimension 2 represents it.

Harp Style

Harp models with two design cases of girders and pylons are simulated in **Fig.23** showing the deflection and stresses due to vibrations induced from natural frequencies of the system.

Dimension 1   Dimension 2

Fig. 23 Mode Shape 1 of Harp Style (two dimensions cases)

The deflections and the stresses are with significant values in the dimension 2 with the weak design of girders and pylons compared to dimension 1 with proper design.

Table.10 Deflection of Stay Cables at Midpoints Harp Style

Point	Deflection M Roller dim 1	Deflection M Roller dim 2
81	0.1623182600	0.3174297200
32	0.1623192000	0.3171216200
4	0.5071626300	0.7218996300
15	0.5067257900	0.7221155800
171	0.0338540380	0.1286343200
91	0.0338524020	0.1286497700
75	1.0029435000	1.0004836000
157	1.0029895000	1.0004784000

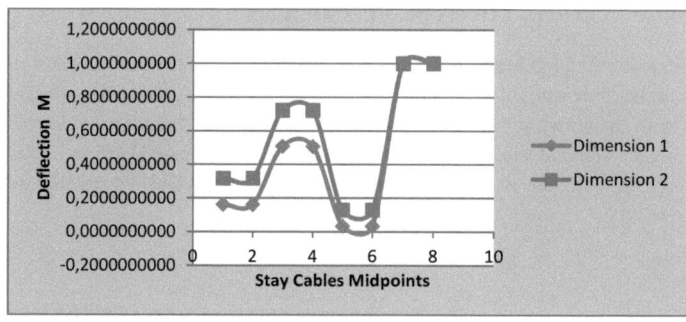

Fig. 24 Deflection of Stay Cables Midpoints Harp Style (two dimensions cases)

Dimension 1 with proper design of girders and pylons has lower deflections of stay cables midpoints of the longest ones.

Table.11 Deflection of Pylons Top Points Harp Style

Point	Deflection M Roller dim 1	Deflection M Roller dim 2
82	0.0001952075	0.1456614000
31	0.0001952070	0.1456616500
5	0.0017857398	0.6017814900
16	0.0017857377	0.6017811900

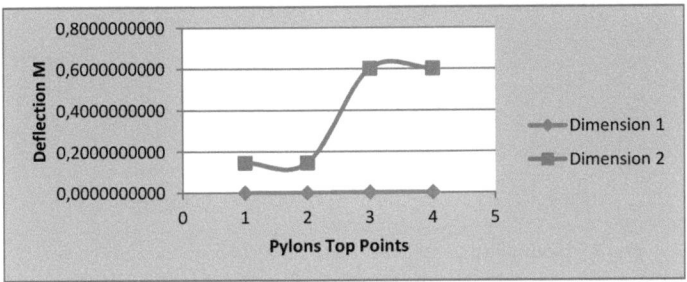

Fig. 25 Deflection of Pylons Top points Harp Style (two dimensions cases)

In the same way the deflection of the pylons top points are lower in the dimension 1 case compared to dimension 2 with weak design of girders and pylons.

Table.12 Deflection of Deck Points Harp Style

Point	Deflection M Roller dim 1	Deflection M Roller dim 2
174	0.0000143549	0.0483334060
24	0.0004605632	0.3044139400
40	0.0018833277	0.6386286600

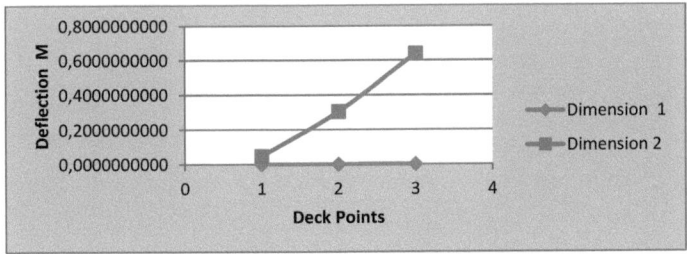

Fig. 26 Deflection of Deck Points Harp Style (two dimensions cases)

Deflection of three points in the left, middle and right of the deck in dimension 1 case is much less than dimension 2 with weak design of girders and pylons.

Semi Harp Style

Semi Harp models with two design cases of girders and pylons are simulated in **Fig.27** showing the deflection and stresses due to vibrations induced from natural frequencies of the system.

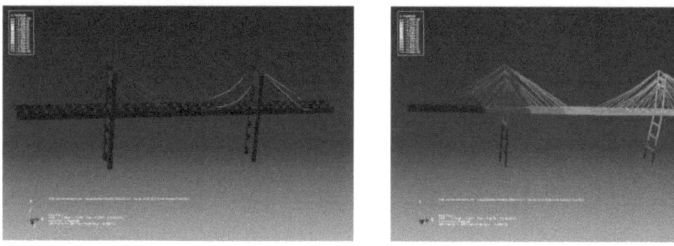

Dimension 1   Dimension 2

Fig. 27 Mode Shape 1 of Semi Harp Style (two dimensions cases)

The vibration and the stresses are with significant values in the dimension 2 with the weak design of girders and pylons compared to dimension 1 with proper design.

Table.13 Deflection of Stay Cables at Midpoints Semi Harp Style

Point	Deflection M Roller dim 1	Deflection M Roller dim 2
101	0.1595814200	0.3189967300
93	0.1595860600	0.3186914900
167	0.4999485900	0.7207176700
185	0.4995139200	0.7209343900
109	0.0320051390	0.1267882000
104	0.0320031340	0.1268004800
293	1.0030704000	1.0004964000
270	1.0031163000	1.0004916000

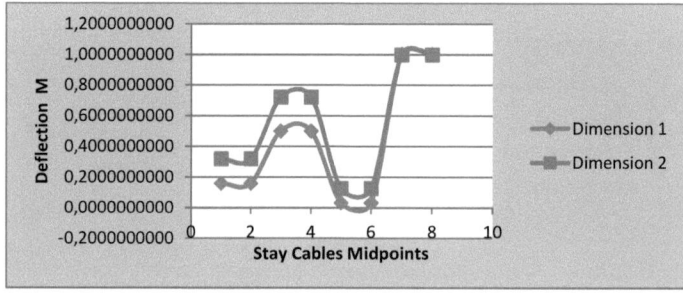

Fig. 28 Deflection of Stay Cables Midpoints Semi Harp Style (two dimensions cases)

Dimension 1 with proper design of girders and pylons has lower deflections of stay cables midpoints of the longest ones.

Table.14 Deflection of Pylons Top Points Semi Harp Style

Point	Deflection M Roller dim 1	Deflection M Roller dim 2
100	0.0001903994	0.1467817100
26	0.0001903988	0.1467820300
166	0.0017720528	0.6047496800
159	0.0017720510	0.6047492000

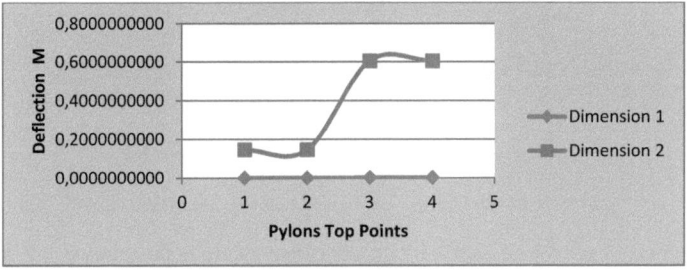

Fig. 29 Deflection of Pylons Top points Semi Harp Style (two dimensions cases)

In the same way the deflection of the pylons top points are lower in the dimension 1 case compared to dimension 2 with weak design of girders and pylons.

Table.15 Deflection of Deck Points Semi Harp Style

Point	Deflection M Roller dim 1	Deflection M Roller dim 2
22	0.0000131256	0.0485976640
171	0.0004570060	0.3057576700
208	0.0018933966	0.6408352900

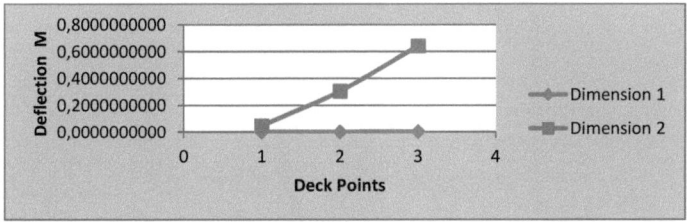

Fig. 30 Deflection of Deck Points Semi Harp Style (two dimensions cases)

Deflection of three points in the left, middle and right of the deck in dimension 1 case is much less than dimension 2 with weak design of girders and pylons.

Fan Style

Fan models with two design cases of girders and pylons are simulated in **Fig.31** showing the deflection and stresses due to vibrations induced from natural frequencies of the system.

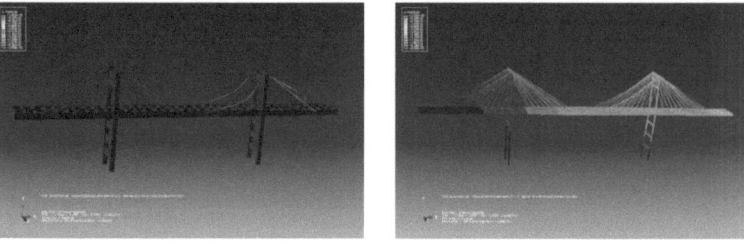

Dimension 1　　　　　　　　　　Dimension 2

Fig. 31 Mode Shape 1 of Fan Style (two dimension cases)

The vibration and the stresses are with significant values in the dimension 2 with the weak design of girders and pylons compared to dimension 1 with proper design.

Table.16 Deflection of Stay Cables at Midpoints Fan Style

Point	Deflection M Roller dim 1	Deflection M Roller dim 2
207	0.1561218400	0.3643554700
202	0.1561622300	0.3641587800
74	0.5017637000	0.7508571100
284	0.5013790700	0.7509700700
235	0.0309473700	0.1570670600
268	0.0309603530	0.1571147300
14	1.0032409000	1.0002950000
26	1.0033560000	1.0002940000

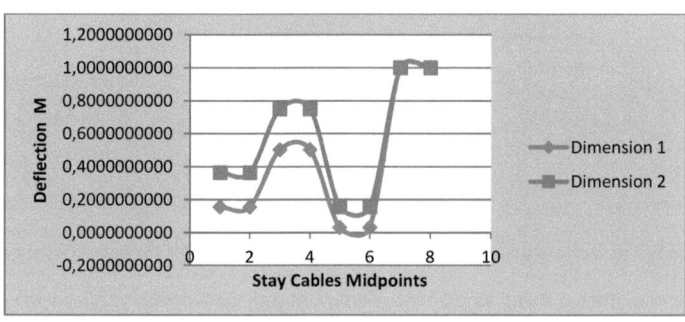

Fig. 32 Deflection of Stay Cables Midpoints Fan Style (two dimensions cases)

Table.17 Deflection of Pylons Top Points Fan Style

Point	Deflection M Roller dim 1	Deflection M Roller dim 2
140	0.0001881050	0.1675071300
132	0.0001881038	0.1675093000
13	0.0018174824	0.5819084600
25	0.0018174590	0.5819063200

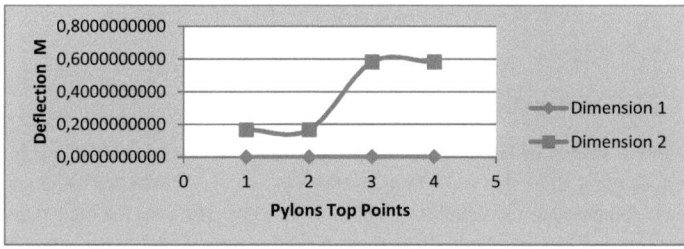

Fig. 33 Deflection of Pylons Top points Fan Style (two dimensions cases)

In the same way the deflection of the pylons top points are lower in the dimension 1 case compared to dimension 2 with weak design of girders and pylons.

Table.18 Deflection of Deck Points Fan Style

Point	Deflection M Roller dim 1	Deflection M Roller dim 2
175	0.0000127711	0.0572492070
125	0.0004518437	0.3181748100
29	0.0018994105	0.6004418100

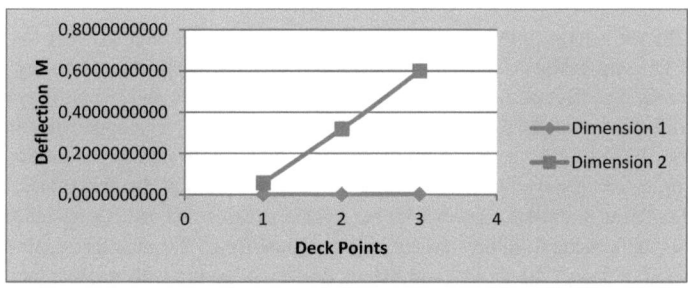

Fig. 34 Deflection of Deck Points Fan Style (two dimensions cases)

Deflection of three points in the left, middle and right of the deck in dimension 1 case is much less than dimension 2 with weak design of girders and pylons.

Conclusions and recommendations

Applying frequency linear perturbation step in ABAQUS to study and analyze the vibration and mode shapes of a cable stayed bridge models imposes assignation of structural parameters change such as the Stay Cables styles, Pylons and Girders cross sections, by examining the response of the global model with respect to the first 10 mode shapes and the response of the Stay Cables, Pylons and the Deck due to the vibration induced from the natural frequencies of the cable stayed bridge system we can list conclusions as:

1- Stay Cables and Boundary Conditions:

The 10 mode shapes of Harp, Semi Harp and Fan styles results with roller, hinged and fixed supports show that the vibration effect on the stay cables for roller support is very low compared to the vibration induced in the stay cables for the hinged and the fixed supports and can be seen in the deformed shape of the model of each case that is the stay cables near to the roller support is only vibrated with a small amount on the stay cables in the other direction , while in the hinged and fixed supports, the stay cables in the two directions are vibrated especially the longer stay cables , and also the natural frequencies for the mode shapes are showing similarities in the values almost for all boundary conditions cases but the mode shapes 1, 2 and 3 which show that the natural frequencies in hinged and fixed cases are higher than roller case , this means that the roller support boundary condition is much prone or near to vibrate due to the first three natural frequencies of the cable stayed bridge system. The mode shapes 7, 8, 9 and 10 are showing the vibration of the stay cables (longest) in all styles. To avoid vibration of stay cables in each case cross ties can be used in the areas of stay cables that are much prone to vibration which means to increase the natural frequency of the system so that to vibration in the early stages.

2- Girders and Pylons

Due to the early stage (mode shape 1) of vibration in the roller support case for each style of the stay cables, changing the dimension of the girders and pylons are with very significant effect on the vibration of the stay cables, deck and pylons especially in the direction near the roller support which can be seen very apparently in the deformed shape of each case, which means that the weak design of the girders and pylons in the cable stayed bridges will lead to early vibration of the stay cables, deck and pylons near the roller support with significant stresses that can be the additional reason of the structural failure. To enhance the stability of this structure against the vibration is to design the girders and pylons due to codes and with suitable factor of safety. Cross ties can be used in addition to suitable damping in the girders that are connected to the stay cables to minimize the mitigation of the vibration induced to poor design of these main structural parameters. Important recommendations for future research works can be pointed as:

1- Using different shapes of pylons and girders to search the effect of these structural parameters on the vibration of the structure.

2- Changing the ratio of damping at the girders and the stay cables connections to discover the suitable damping ratio that can be used to minimize the mitigation of the stay cables vibration in all cases.

3- Using cross ties in the models to distinguish the difference between situations with and without cross ties in relation with the effect on the global vibration of the structure.

References

1- Fujino, Yozo, Siringoringo, Dionysius. Vibration Mechanisms and Controls of Long Span Bridges: A Review. Structural engineering international-Vol. 23, No.3, pp. 248-268 (21) August(2013).

2- Liuchuang Wei, Heming Cheng, Jianyun Li, Modal Analysis of a Cable-stayed Bridge. International Conference on Advances in Computational Modeling and Simulation, Procedia Engineering 31, 481 – 486 (2012).

3- He Zhang, Xu Xie. Dynamic responses of cable-stayed bridges to vehicular loading including the effects of the local vibration of cables, Journal of Zhejiang University-Science A (Applied Physics & Engineering), 12(8):593-604(2011).

4- Limin Sun, Hongwei Huang, Yonglong He, Yagang Zhou. Simulation on cross ties for vibration control of long span cable stayed bridges. Proceedings of the international conference on computing in civil and building engineering-The University of Nottingham, UK(2010).

5- G. E. Valdebenito, A.C. Aparicio. Dynamic characterization of cable stayed bridges: a comparative analysis. Fifth international conference of seismology and earthquake engineering, Tehran-Iran, May (2007).

6-Yuh Yi Lin, Yen Lung Lieu. Geometrically nonlinear analysis of cable stayed bridges subject to wind excitations, Journal of the Chinese Institute of Engineers, Vol. 26, No. 4, pp. 503-511 (2003).

7- J.M.Ko, Z. G. Sun, Y. Q. Ni. Modal analysis of cable stayed Kap Shui Mun Bridge taking cable local vibration into consideration. International conference, Advances in structural dynamics, 1, 529-536, (2000).

8- I.W. Lee, D.O. Kim, G.H. Jung. Natural frequency and mode shape sensitivities of damped systems: part2 multiple natural frequencies. Journal of Sound and Vibration 223(3), 413-424,(1999).

9- U.Starossek. Cable dynamics- a review, Structural Engineering International, Vol. 4, No 3, pp.171-176 (6), 1 August (1994).